Ideas for Technology Integration for Teachers

by Mark Page-Botelho

Published by Mark Page-Botelho
ISBN 978-0-6151-9324-3
Email: mpage@netpb.com

Table of Contents

Introduction

Incorporating technology can be a difficult process if one is not familiar with how to integrate software and hardware into their curriculum. We have all had opportunities to learn ways in which to incorporate technology, but many of those were too difficult and you may not have felt capable enough to try it in your own classroom. If teachers were to incorporate just one aspect of technology, it is far better than not incorporating any at all.

The Autobiography and Video projects, on which this book is based , are an easy way to integrate technology. It is important to keep in mind not to go all out and try many different new things at once. Take it slow, even incorporating one piece of technology a year, is a step in the right direction.

Benefits

Using technology has many benefits especially for those students who are visual learners, or have trouble organizing their thoughts. Other benefits includes making edit and revisions easier for students, and storage of work in digital portfolios to be passed on from year to year. Lastly, students will become more familiar with the tools they will need to use once in college and the work force.

Strategy

One strategy to insure that one does not feel over-whelmed, is to use one meaningful lesson. Become proficient at one thing rather than try to use many different types of technology, which can confuse both teacher and students. Practice doesn't necessarily make perfect, but it can make one proficient.

Where to Find Help

For those teachers who are not comfortable using technology, you can usually find help at the school or district level. Use your school's technology coordinator or computer lab teacher for help. Also, having peers who are familiar with the software and hardware at your school are a great resource. Many of which may already have lessons planned using technology that they might be willing to share with you.

Another great resource for find help and tutorials for both students and teachers is the Internet. I've been working in the technology field for over a decade and everything I've learned has been through self discovery with a lot of help from information found on the Internet. The simplest method is to type a question in Google as if you were asking a knowledgeable person. For example, "How do I change the font in Open Office?" With a little hunting and pecking through the results, you'll more than likely find your answer.

Autobiography Project Overview

The purpose of the lesson is to facilitate the writing process using technology tools, such as Kidspiration or Freemind mind mapping software, and word processing software, such as Microsoft Word or Open Office. Project utilizing technology can instill a sense of meaningfulness for students. Also, it is our responsibility as teachers to prepare our students for a highly competitive work force by fostering their competence regarding technology, and using it as often as we can to help them become proficient at it.

It is our responsibility as teaches to help foster our students into "Competent Digital Users," those who have the ability to acquire the ability to use technology by intuition learned form experience. Many people believe that children are just better at technology than adults. This is partially true. Children who use technology are good at what they know, but those who don't use it or have a narrow technological focus are not. If a child spends 20 plus hours a week on a gaming console, they will be very skilled at that one particular piece of technology. This however does not mean that they are knowledgeable or skilled at using a computer workstation, or how to use a word processor to its full advantage.

Some students may even be at a disadvantage by how they use technology. For example, a students who does not learn how to use a search engine efficiently may be at a disadvantaged when completing projects which require Internet based searches, as they will waste time in needless searches.

As with every skill we learn, the more you do it the better you get at it. How often one uses and what types of technology will affect their expertise. If students use technology in their everyday lives at school, they will become more efficient at it. Starting with simple skills in elementary, will allow students to progress as they advance through the grade levels. If all teachers participate, by the time students reach graduation, they will have mastered and be prepared for college or the workforce.

Technology and the Writing Process

This project demonstrates an example integrating technology and writing. The *National Educational Technology Standards* (NETS) and Language Arts standards based on the 6+1 Traits of Writing, produced by the *Northwest Regional Educational Laboratory* in Portland, Oregon have been used for planning and assessment.

Using a Mini-Quest

Using a static web page, a web page that never changes, for listing the steps a student has to follow to complete the project is a great way to introduce them to a mini-quest or web-quest. A web-quest is a web page that is a story or script based inquiry based lesson for students. A mini-quest is more informal and not script or story based. It is still inquiry based, but more slimmed down and is usually quicker to do that a full web quest.

Mini-Quest Example

Autobiography (W1, W2, & W7)

Lesson Summary:
Students will demonstrate their ability to write a narrative essay which includes the 6+1 elements taught by writing a 5 paragraph autobiography.

Problem:
You are the mayor of a small town. The people in your town want to dedicate the front page of their newspaper to you for all the hard work you've done for the town. The problem is, since the only town reporter is sick at home with the Flu, you have the write the article.

Process:

1. make a web in Kidspiration about you. You will need at least 3 B ideas.
2. convert your web to an outline.
3. export your outline to open office.
4. Fill in the details for your 3 B paragraphs.
5. Include an introduction paragraph
6. Include a conclusion paragraph (hint: use your introduction paragraph, just change it a little bit)
7. Ask if you can print the final copy.

Assessment:

Five paragraph essay, with one photograph insert. Included in the assessment:
Using the 6+1 rubric students will be graded on Ideas, Organization, and the writing process.
The essay must included an Introduction paragraph, a minimum of three detail paragraphs, and
a Conclusion paragraph.

Once familiar with using a web page to give instructions, creating mini-quests or web quests is a great technology tool. It can be very dynamic and allows students to use their judgment on how to solve a problem or research. Thereby instilling a sense of internal motivation. Another benefit for posting an assignment in a mini-quest is that it is available at all time and has relevant links to help students stay on track. It also makes sharing your work with others.

Kidspiration and the open source alternative Freemind, are mind mapping programs. They allow students to brainstorm and organize their thoughts before sitting down to write. Kidspiration has the advantage that it allow students to export their final mind map in an outline format to a word processor. This saves time, so the students don;'t have to copy an paste their ideas into a word processing document.

Lesson Summary

Students will demonstrate their ability to write a narrative essay which includes the 6+1 elements taught by writing a 5 paragraph autobiography.

Process

Introduce the writing project using a Mini-quest. This gives a student a purpose to help internalize their motivation. The great part of using a miniquest is that you can create a story in which the students plays a part that they get to chose.

Kidspiration Mind Map Example

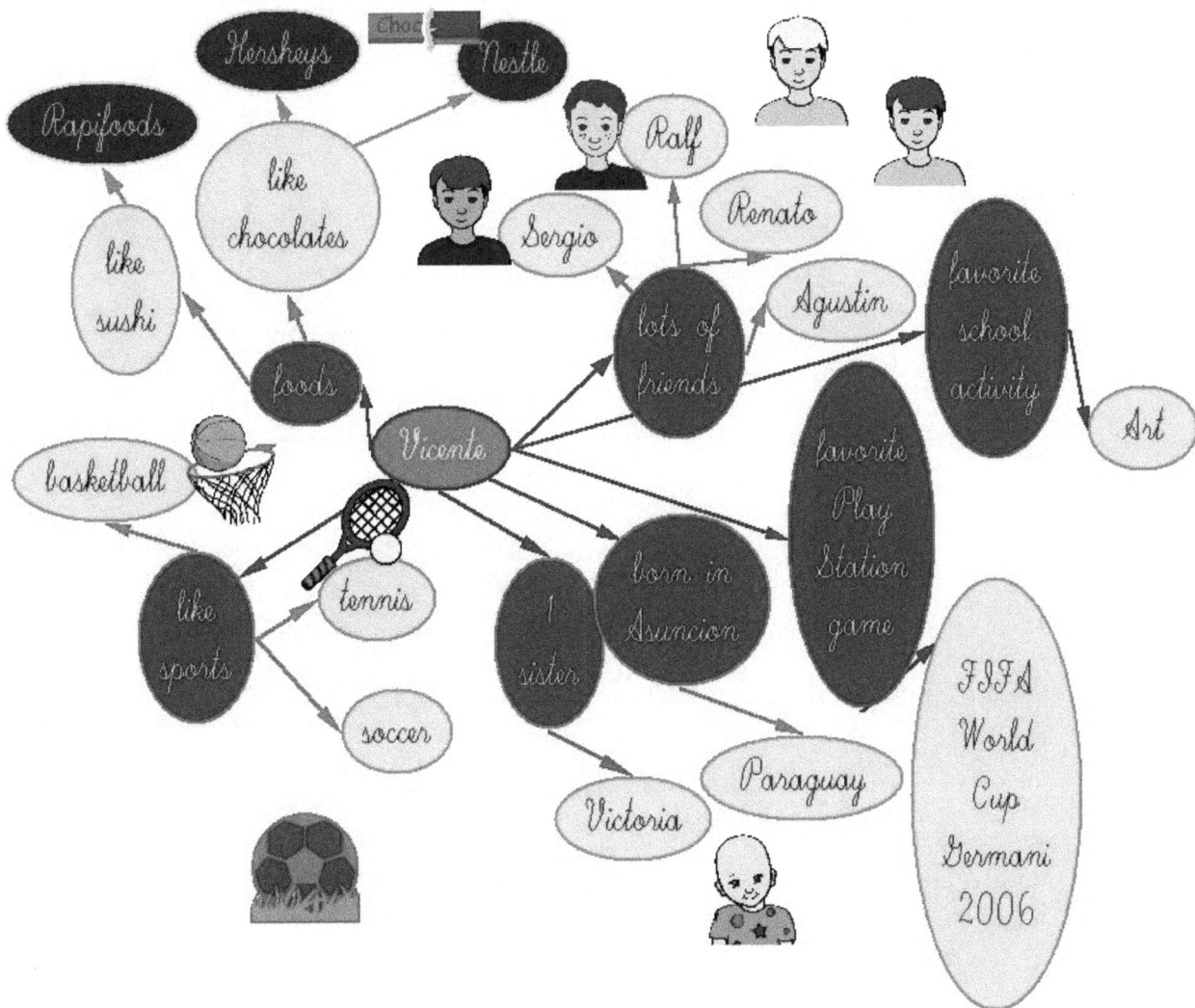

Use Kidspiration for Brainstorming - Using Kidspiration or its open source equivalent Freemind for brainstorming out ideas in a visual format. Using software for mapping out ideas is much more efficient and less frustrating fro students, who can easily manipulate their ideas using software rather than paper and pencils. Using software also has the added benefit of allowing students to use graphic images for each item in their map, which allows them to easily categorize their thoughts. Students should spend a good deal of time on this step as it will save a lot of time and trouble during the revision processes.

Kidspiration Outline Example

Luciana.kid

File Edit Goodies Sound Teacher Help

Luciana

My favorite sport

basketball

tennis

My pet

Leon

Aron

Times New Roman   14   A A A A   I.II.

Start   02:35 p.m.

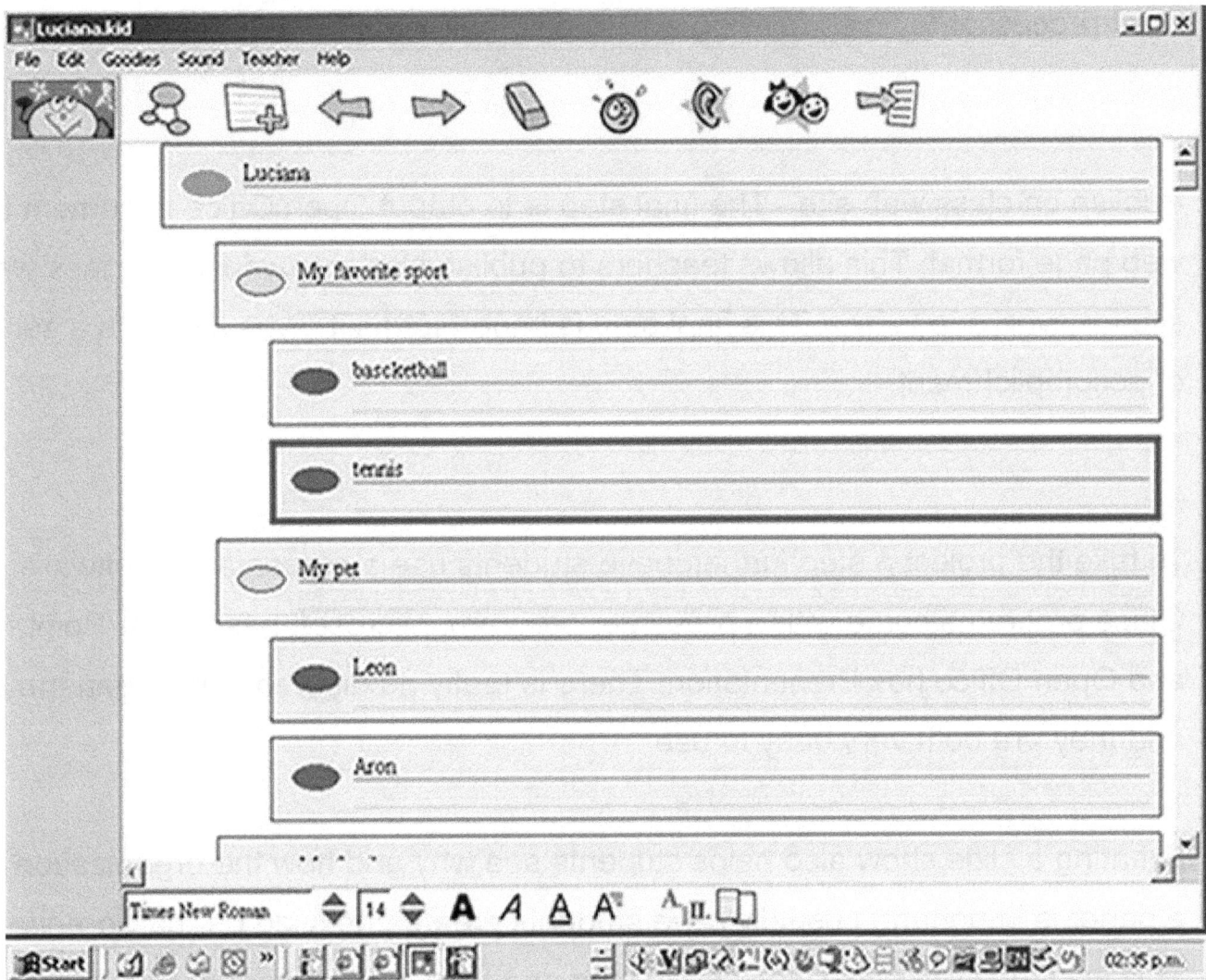

Another task that should be done in the brainstorming step is to write notes or thoughts about each of the items in the mind map. This helps students to come up with details about the items they want to use and will later help them remember what they were thinking about during the brainstorming sessions.

Create outline form in Kidspiration - converting a graphic organizer to a linear outline form gives students a chance to verify their sequencing and to further revise their work. Freemind does not have this option, but one can still create an outline by copying and pasting the items into a word processor such as Word or Writer.

Outline Output to Word Processor to revise and edit – Have students use a word processor to finish revising and to edit their work.

Publish on class web site - The final step is to output Open Office document to a web page format. This allows teachers to publish students work on a class web site for everyone to see. This final step is important as it gives students a sense of accomplishment.

To take the project a step further, have students use a presentation software. All Office programs come with a presenter program. Microsoft has Power Point, and Open Office has Presentation. There is really no difference between the two and they are both very easy to use.

Creating a slide show also helps students see why and how the organization of a paper is important. I usually have students create one slide per paragraph. This way the slide show won't be to long and they see the importance of paragraphing.

To help students stay on track I do not allow them to play with fonts or styles while initially creating the slides. I do however have them use royalty free photo web sites to find images that fit the idea of each individual slide in their presentation. It would be helpful to list the chosen slides in the miniquest.

Once the slides have been completed I would then allow the students to play with fonts and colors, but I would give them a limited choice, otherwise there is a good possibility that they'll make the slides difficult to read.

Language Arts Writing Standards

Understands and uses the writing process effectively: All teachers will grade this differently. I look to see if students are creating the different drafts, and that they are making corrections that I point out and fixing mistakes or making revisions on their own. Using electronic word processors are great for viewing revisions. They all have a version checker. Just make sure that included in your writing process, you have students use a naming convention which insures all revised documents are named differently. When newer versions are named differently, it makes it much easier to go back to a previous version if necessary. This would be a very cumbersome task if done with paper and pen.

Uses grammatical and mechanical conventions correctly: Using spelling and grammar checks in a word processor is a great tool. Think of it not as cheating, but rather mini lessons. Students will notice their misspellings as they will be pointed out to them by the computer. They will also have to check among the possible correct spellings. Students will also over time, realize that some words that are suggested by the computers are not the right words that they had in mind.

Clearly expresses ideas in writing: Having the ability to edit and revise electronically will make the writing process much more enjoyable for your students. In turn it should alleviate the apprehension that many students have about writing and how time consuming it is when dealing with edits and revisions. It is much more time consuming and frustrating to have to use paper and pen. Just think about if you had to write all your curriculum documents out by hand!

Word choice is varied, interesting, and descriptive: Using a word processor for writing is a great way for students to find new words to use. Show them how to use a thesaurus and watch their vocabulary grow.

Uses voice which grabs the reader's attention, shares feelings, and considers audience: I like to have students write from a perspective of one character. This is a great lesson to help students understand what Voice is in writing. With electronic documents it is very easy to share with a classmate ones work. I usually have students pair up with a classmate and have them read the perspective of the student who did the writing, allowing the students to see how perspective changes a story.

Sentences vary in length and style: Having the ability to cut and paste makes the revision process easier. When considering style and length of sentences, demonstrate to your students how sentences can be moved around and how that might make the idea of the paragraph change to become more clear.

Uses topic sentences, supporting details, and closing sentences to organize writing: Using a mind mapping software makes the organization of a paper more structured. The more time spent looking at their mind maps, the more organized their papers will become. A great exercise would be to have students write a short essay first not using mind mapping software. Then have them write another essay using it. This will clearly show them the benefits of using software to organize their thoughts.

Based on the North West Regional Educational Laboratory Writing Traits Model

Technology Standards

Finding good technology standards to use for a school can be difficult. There are basically two thoughts on the matter. First, are the school standards to be based on skills, such as opening and saving document, typing speed, or ability to modify an image. Secondly, is the school more interested in using technology in a meaningful project based way.

The first school I taught at, the elementary based their standards primarily on skills, such as typing and basic computer operation. I decided that it was more meaningful to incorporate those skills in a meaningful project, which is where the *Arizona State Technology Standards* fit in.

Below are listed the skills used for the Autobiography Project.

- Understand Basic technology operations and concepts, Students demonstrate a sound understanding of the nature and operation of technology systems.
- Open programs
- Open/save files from/in the network
- Use correct sitting, hand, and arm positions and fingering to type a

document employing the full alphabetic keyboard.

- Uses technology tools to enhance learning, Create a multiple page document using paragraph, page formating and graphic features
- Choose font, size and style
- Use space bar and return keys appropriately
- Use the tab key to indent
- Use the undo feature of the application
- Choose text justification
- Use the cut/copy/paste feature of the application
- Set page border and number
- Insert images
- Resize images

Writing Criteria

The 6+1 Writing Traits rubric used for this project was based on the *North West Regional Educational Laboratory* Writing Traits Model. The five point rubric was slightly modified to fit the needs of our school. We also tried to limit the amount of things we were looking for in order to make the 6+1 Traits easily remembered by the students.

Ideas

- I added interesting details.
- My paper is about one main idea.
- Details present but not precise. Reasonably clear ideas.

- Paper appears to lack adequate details and the paper loses focus.

Organization

- I wrote my topic sentences correctly. (A sentences)
- I correctly added sentences with supporting details. (B sentences)
- I correctly added sentences with supporting sub-details. (C sentences).
- I have a clear beginning, middle, and end.
- Paper does not have a clear beginning, middle, or end. Some of the detail structure is not complete.
- Paper does not have a clear beginning, middle, and end. Detail structure is not complete.

Voice

- My writing is interesting and uses details and words that only I would use.
- My writing sounds right for I express feelings and emotions in my writing.
- Whoever reads this story paper will feel happiness, sadness, fear or another emotion.
- I feel confident about reading my story or paper out loud to others.
- Pleasing, yet safe. Writer/Reader connection fades in and out. Purpose inconsistent.
- Flat, No purpose.

Word Choice

- I correctly used words.
- I used uncommon words.
- I didn't use too many overused words like nice, good, bad, pretty, or ugly.
- Some words are misused.
- Simple words used inappropriately. All words are common.

Fluency

- I correctly used transition words within my paragraphs. (first, also, next, because)
- My paragraphs sound good when read aloud and are easy to understand.
- I always use complete sentences.
- I mix longer, complex sentences with shorter ones.
- Attempts compound and complex sentences.
- Choppy, rambling, repetitive.

Conventions

- I used capitals at the beginning of my sentences and with proper nouns.
- I indented at the beginning of each paragraph.
- Each sentence ends with correct punctuation.
- I have spelled almost all of my words correctly.
- I correctly used verbs in past and present tense.
- Spelling generally correct, Routine paragraphing, grammar, capitalization.
- Spelling errors impede readability, Many grammar errors, lack of paragraphing.

Instead of using the +1, which signifies presentation for the 6+1 Writing Traits, our school substituted it for the writing process. Below is what we look for while grading the writing process.

Writing Process

- Evidence of pre-write
- Completed rough draft
- Evidence of revision
- Evidence of edit
- Final copy turned in

- One to two steps missing.
- More than three steps were not completed

Technology Rubrics

Nets

Standard1: Understand Basic technology operations and concepts

-Students demonstrate a sound understanding of the nature and operation of technology systems.

Autobiography Checklist

1. Initials and class are at the upper left hand side
2. Title is centered
3. Title font in color
4. Title is in bold
5. Title font is between 14 and 18
6. Title font is different from Times New Roman (clear and easy to understand)
7. All first sentences of paragraphs are indented
8. The paragraphs font size is 12 or 13
9. The paragraphs font is Times New Roman
10. The line spacing is double
11. The page has a border

12.   The page has page number

13.   The page has the author's picture centered on top of the page

Standard 3: Uses technology tools to enhance learning

-Creates a multiple page document using paragraph, page formating and graphic features

Standard 1: Locate and open applications, such as Kidspiraton and Open Office Writer

-Open/save files from/ in class folder

-Use home row on the keyboard to type

-Export files from Kidspiration to Open Office

Essential Question

How can we live, learn and work successfully and responsibly in an increasingly complex, technology-driven society?

Arizona Department of Education Technology Standards

STANDARD 1: **Fundamental Operations and Concepts**

Students understand the operations and function of technology systems and are proficient in the use of technology.

STANDARD 2: **Social, Ethical and Human Issues**

Students understand the social, ethical and human issues related to using technology in their daily lives and demonstrate responsible use of technology systems, information and software.

STANDARD 3: Technology Productivity Tools

Students use technology tools to enhance learning, to increase productivity and creativity and to construct technology-enhanced models, prepare publications and produce other creative works.

STANDARD 4: Technology Communications Tools

Building on productivity tools, students will collaborate, publish, and interact with peers, experts and other audiences using telecommunications and media.

STANDARD 5: Technology Research Tools

Students utilize technology-based research tools to locate and collect information pertinent to the task, as well as evaluate and analyze information from a variety of sources.

STANDARD 6: Technology as a Tool for Problem Solving and Decision making

Students use technology to make and support decisions in the process of solving real world problems.

Wrapping Up

Once we are comfortable with ourselves and our teaching curriculum, we are not very apt to change our ways or try something new. As educators it is our duty to better ourselves so that we can best teach our students and prepare them for the future. It is safe to say that technology will be a big part of our students lives, therefore we must help them learn the necessary skills they will use when they enter the work force.

As I stated earlier, it is better to make one small step forward, then to make no progress at all. Learning one new way to incorporate technology into ones curriculum should be an easy feat for any of us. Incorporating technology into a project that you already do, will make integration easy and this will show in your comfort level and the students will also be more relaxed and trusting when they notice the teacher at ease with a lesson.

Remember that if you get stuck or feel unsure of what to do next, talk to colleagues at your school, or even ask for help at the district level. All schools have technology help desk or knowledgeable staff that would be more than willing to help.

Where now?

Coming up with a solid lesson that can be easily modified and is easily implemented in the classroom is easy. The idea is to come up with a lesson that you will use often so both your students and yourself will become familiar enough with the technology that the skills learned will be used often and increase in proficiency. Use Freemind or Kidspiration to help write an autobiography, then create a digital presentation.

Video Project Overview

This project was designed to integrate video production with writing into a real world project. The goal was to give students a project that they would feel proud of and give them a final product that they could share with family and friends. Also, the project helped out the school by taking a school wide issue and making something that they could use to help them in its resolution.

Hardware / Software Requirements

Simplified 6+1

By far one of the best writing assessments I have seen is the <u>6+1 Traits of Writing</u>, produced by the *Northwest Regional Educational Laboratory* in Portland, Oregon. There was a slight modification in the language in their rubrics just to help clarify what was being looked at for the sake of early elementary grades.

Multiple Intelligences

In order to insure that all students have an equal chance of learning, the Howard Gardner Multiple Intelligences were incorporated where feasible. Even though not all of the multiple intelligences were taken into account perhaps other may find a way to fit them in.

Bring in other Professionals

Another aspect of the project included getting help from others at the school. This was to get the help and expertise of individuals such as the counselor, second language teachers, technology specialist, PE teachers, or parents. The wealth of knowledge that can be garnered from others is a great resource that should be used when it can. It makes the whole overall experience much more rewarding and also allows students to get different perspectives that they may not have otherwise have gotten.

Pre-Writing Stage / Writing Stage / Post-Writing Stage

The writing part of the project relies on the writing process. Students are taught and practice the writing traits for all writing. It is important to keep revisions of the student work. It is a great tool to use to show students and their parents the progress that their children are making.

Hurdles

This is a complex and time consuming project. There are things to watch out for and tricks that I learned the hard way that will be discussed in order to make your experience more fruitful.

Software

There is essentially only two unique pieces of software which are needed to do the movie project. The first is Microsoft Movie Maker (iMovie for Mac). This is a free program and is very easy to use. If your video camera is digital and has a firewire connection and so does your PC, you can easily use Movie Maker to capture video. Even if you are using an older video camera it is still an easy chore, just not as fast.

The other essential piece of software is KidSpiration (Freemind is a free open source alternative). Kidspiration allows students to create a mind map, otherwise know as brainstorming. I like using Kidspiration, because it allows students to easily manipulate their thoughts and re-arrange them as they are putting together their thoughts. It also allows them to create an outline view, which helps in their organization of their paper. The outline is then easily exported to a Rich Text format, which all office programs accept.

Hardware

The main component that is needed to make a movie is a video recorder. Obviously when most people think of a video camera, they think of the traditional camera that uses some kind of tape and is usually bulky. Today there are many more options available. Depending on the desired output, one can use a traditional video camera (tape) or a newer smaller hard drive or flash video recorder (very expensive). I have even had a class use the video recorders in their cell phones.

The more money you invest in a camera the better the quality and easy ability to transfer and edit the video itself. The higher end cameras can create a fantastic quality video on par with DVD and high definition playback. Cell phone video recorders are on the other end of the spectrum, and the video is really only good enough for playback on a computer in a very small viewing window.

Other equipment that is useful to have, but not necessary is a tripod, which can easily be substituted by a shoulder of another student. Lighting is also a great extra to have if where you are shooting the video does not have enough ambient light. Timing your shoot if possible can alleviate the need of extra lighting, or use a stage where the lighting is good.

Finally a good fast computer with plenty of RAM is helpful. Video editing can be done on an older slow computer, but the editing of the video can be tedious and take up valuable class time.

6+1 Writing Traits

It was our belief that we didn't want the 3rd graders to stress out over the many facets of the 6+1 writing traits so we decided to simplify them. In doing so the students we believed would have more time to work on the video project as a whole and would easily be able to remember the main idea of each of the traits which we believed to be more important at this stage in their learning. Below are the main idea that we wanted our students to come away with after the project was completed.

IDEAS

My writing is about ONE idea.

ORGANIZATION

I have a good introduction and conclusion. I have a clear beginning, middle, and end.

VOICE

My writing has a clear point of view, and I used "feeling" words.

WORD CHOICE

I used powerful, interesting, and engaging words. The words I used are a good choice to describe my topic. I used figurative language.

SENTENCE FLUENCY

I used compound or complex sentences. All my sentences start differently.

CONVENTIONS

I checked my spelling and grammar for mistakes. All my paragraphs follow the rules of a good paragraph.

Grading

Grading can be a touchy subject depending on who you talk to. I found there to be essentially two schools of thought concerning group projects. First, group work is important as students help each other learn new ways of doing things and each student brings their own perspective which one teacher can not do alone. Secondly, group projects are ineffective as students have a hard time coordinating their efforts and usually one or two students do the work of the group. Thereby students are not fairly graded.

In order to gain the benefit of a group project yet still retain accountability for work done, I combined the two schools of thought into one. I had all assignments done individually first. Then as a group the students would pick the best version or pieces of different versions, and make the revision draft together. This way the students learn the lesson at hand and are graded individually, and get the benefit of group cooperation.

Note: All writing assignments were given individually first, then group decided and re-wrote as a group (two grades, individual and group)

Multiple Intelligences

The Howard Gardner's Multiple Intelligences include seven areas that he describes. They are Linguistic, Logical-mathematical, Musical, Bodily-kinesthetic, Spatial, Inter-personal, and Intra-personal intelligences. I decide on using only four of the seven. If one researched the different intelligences you may be able to incorporate additional ones somehow. Using the multiple intelligences, helps make projects more dynamic, and insures that more students will walk away from the project successfully learning.

Teachers and Specialists Working Together

One aspect of the video project that is exciting, but also takes the most effort to coordinate is the inclusion of others teachers, parents or specialists. Regardless of the theme the people around us have a great deal of knowledge that is a wonderful resource. By included others in the project you also open up different perspectives that may not otherwise get covered, or covered well. The students will only benefit from the inclusion of others.

A class or school wide theme is a place that one might look when designing the project. I like to have a social constructionist theme, as the finished project would lend itself well to use in an assembly for the benefit of other students. Also, if used as a real product, the students will achieve a higher sense of accomplishment.

Project Ideas

An example of themes I have used for a video project are listed below.

Bullying, this project was done in elementary using the school wide anti-bullying strategies taught by the elementary counselor.

Cyberbullying, this project was done in high school and the information on the harm and strategies to fight it were gained through research done by the students.

Country Information, this project was also done in high school and was for the benefit of the Director General for use at the many job fairs he attends. The information was used for potential teachers to learn about the host country.

Incorporating Specialists Examples

Below are some examples on how one can incorporate others into a project.

Art – Have students create props for movie during an art class.

Computers – Have technology instructor help students learn writing and brainstorming programs.

Counselor – Explanation of tough issues is a great way to incorporate counselors. They can give insightful strategies as well on how to deal with many issues that students can incorporate into their projects.

Language – If your school has a foreign language specialist, have them help with a project by adding a foreign language aspect, such as using equivalent words for specific items in another language.

Parents – Many parents have expertise in the theme you will be tackling. Invite them to help participate by giving a presentation. They will be thrilled to help their child's class!

Pre-Writing Stage

The pre-writing stage consists of three steps, and its main purpose is to give students enough background knowledge to successfully complete the project. The first step involves introducing the guest speaker, topic, and note taking. For the second step, have students write about the topic at hand. Finally, step three involves reading stories about the subject matter.

The first step, otherwise known as the state objective, is to insure that the students know what the purpose of the project is. Uncertainty is not a good thing for students. They need to know what they are up against and to give them time early on to start planning.

The second step involving the guest speaker will give students a solid foundation of information to work with. Guest speakers do take some effort to set up, but they are well worth it. Students of all ages enjoy listening to someone other than their teacher. The novelty is enough in itself to capture their imagination. More importantly it insures that students get a more well rounded perspective of the topic at hand. The note taking portion is important for gathering information, but also students need to practice their note taking skills.

The third step give students more background knowledge and allows them in their writing to start putting together their own opinion of the topic. Find other material on the web or even your school library. Let students know where to find it, or better yet, check out as much material as you can and have it on hand in the classroom. I like breaking the students into a few groups, having one use the computer lab, while other read book or work on other parts of their projects.

Writing Stage

Modeling Inappropriate Behavior

Students of all ages get excited at the prospect of making a movie. Many of them when they think of a movie automatically think of action movies that they are accustomed to watch. These movies may contain violence or have sexual connotations that are not appropriate for school based projects. Please stress that the projects including the writing is to contain no violence or sexual overtones. I would recommend telling the students that showing the after effects, such as a student crying and explaining she was bullied, is more than enough to get a message across. Do not let them depict that inappropriate action as that would entail modeling inappropriate behavior.

Below are the six writing traits and examples of activities that can be done for the video project.

IDEAS: Group Work, Draw at least two Mind Maps (Web) of your ideas for a Bullying Video Using Kidspiration

Organization: Story Board, students had to have a problem and solution provided by the Counselor

Word Choice This was a difficult portion of the project for my third grade class. They were second language learners, which presented a unique aspect, as well as the male students had a difficult time conceptualizing how to show a problem without modeling inappropriate behavior. The solution was to give many examples and to provide students with pre-made lists of words (theme based) they could use. Many activities which provided the definitions was used to help students learn the words.

Voice: Have each student pick one character and have them write a paragraph from that point of view. One great exercise was the ripped pants wring prompt. In this lesson, I had students grouped into six different groups. I told them how I was teaching the class unaware that I had a big rip in my pants. Each group was assigned a perspective, such as, the teacher, a student, the pants (always fun), another teacher, a parent, and the principal.

Sentence Fluency: Discuss not starting a sentence with the same word, using conjunctions, and figurative language.

Conventions: T-Script, proper use of quotations.

Post-Writing Stage

Collecting Props and Costumes

This stage of the process takes a long time. If your school has a drama department, I would suggest seeing if you can barrow the props and costumes needed. Even better if time permits, include the art instructor in this process. Have them make the props or costumes. This is an easy way to incorporate other specialists.

Practice Skits

Practicing skits in an already busy day can be a difficult process. If you have an aid it will make this step easier, if not try and get students to get together outside of school, or have a parent volunteer during the day to take individual groups to a private location where they will not disturb others. Making sure that groups have all their work done prior to practicing is a great way to get unorganized students motivated. Students love to go practice their skits, whatever the age.

Edit & Revise

Allow time after the practicing of the skits to allow students to revise their skits. You should encourage them to revise at least one aspect of the movie per practice to help them realize the importance of revising. After you decide that a group has fulfilled your expectations for revising, then have them edit , making sure that their work is publishable. Every teacher has their own perspective on this, I allow some unrefined work to pass, as it adds character and the age of authorship, and keeps children from worrying the technicality of writing and keeps it fun. The older students become the more refined their writing will become almost naturally with age.

Action! Film Commercials

Once you have acquired the props, writing, and you have the equipment, you're ready for shooting. I usually have a couple of dry runs where the students preform as if you are filming, just to insure everything is ready. It also allows students to adjust to the filming procedure. Some things to look out for during this step is make sure there is enough lighting, that students speak clearly and loudly. Most importantly that you give a few seconds of wait time from when you click the record button to when the students start to preform. This makes the editing process much easier.

Editing Video

Depending on the ages of your students will determine how much involvement the students will have in the editing process. For my third grade class, after I trimmed out the irrelevant portions of the video, I had the groups create the introduction, scene changes, conclusion, and credit transitions. I would choose which students get to do what to save a lot of heart ache and trouble with compromising.

The software I used was Microsoft Moviemaker. For Macs, use iMovie. Both are incredibly simple to use and free. Even my third graders had no trouble and easily were able to use the software with little intervention from me.

Show Commercial!

Once you have completed the editing process, invite parents to view your students work. If it was a school wide theme, show your videos during an assembly. Distributing the final product can be expensive and time consuming. DVD's ave come down in price, and perhaps you can find a parent volunteer to help in coping the videos. Another option is to ask your technology coordinator if the video can be formated so it can be streamed live from a password protected portion of the school web site.

Hurdles to Keep in Mind

Time

The movie project is very time consuming. It took me at least one quarter to accomplish the project. Obviously the more time you do it the better and faster you will get at it. But if it's your first time, allow plenty of time and expect that some aspects might have to be done twice.

Working in a Group

Compromise is key! In all classes, group work can be a problem, especially with younger students who are just learning group social behavior. I taught my students a lesson specifically on compromising. I believe that this help immensely and would recommend that you do it as well before starting the project.

Review

Bring in other Professionals (Counselors, Language, Technology, Physical Education Teachers, Counselors, or Parents). The more help you can get, the easier the project will be on you. Do not burn yourself out when there are plenty of colleagues that would love to help.

Using the 6+1 Writing Traits & Writing Process in your project will help your students become better writers. The point of doing the video is to get students to have fun while writing.

Multiple Intelligences will insure that all your students will learn something from the project. It also will make the project more dynamic and interesting for the students.

Group Work / Cooperation, Learning from Peers is an important skill that is a schools shared responsibility with th parents. Students spend a great deal of time at school, sometime they spend more time with teachers than with their parents. Therefore we are just as responsible for teaching social skills.

Technology will be an important aspect and job skill needed by our students when they enter the workforce. Have fun while learning technology is a great want to teach these skills. Some programs used to for the video project include Kidspiration, Video Capture & Editing software.

Real World Project producing something tangible is a great way to motivate students. I have never met a student who was not excited while working on their video.

Helpful Items

Feel free to copy the following pages for use in your classroom.

Elementary 6+1 Check List

IDEAS

My writing is about ONE idea.

ORGANIZATION

I have a good introduction and conclusion. I have a clear beginning, middle, and end.

VOICE

My writing has a clear point of view, and I used "feeling" words.

WORD CHOICE

I used powerful, interesting, and engaging words. The words I used are a good choice to describe my topic. I used figurative language.

SENTENCE FLUENCY

I used compound or complex sentences. All my sentences start differently.

CONVENTIONS

I checked my spelling and grammar for mistakes. All my paragraphs follow the rules of a good paragraph.

Bully Brainstorm

Bullying Brainstorm

Think of different ways that people bully at school. Where does it take place, who is it happening to? For every idea you have, think of a way to solve the problem.

Draw **at least two** Mind Maps (Web) of your ideas for a Bullying Video.

Story Board

Verb Outline

Write the main "verbs" that make your story. Make sure you use verbs. They need to be action words. Make sure you use a complete sentence.

1.

2.

3.

4.

Now draw a picture showing the action of the verb you used for each sentence.

T-Script

Action:	Voice:

T-Script

Action:	Voice:
-Sergio and Jose are passing a soccer ball back and forth.	*-Sergio says to Jose, "We're lucky Mr. Page is nice and gave us extra recess!"*
-Jose kicks the ball to Sergio.	-Jose says, "yes, Mr. Page is the nicest teacher at ASA."
Paulina walks by eating an ice cream.	
-Nicole walks past Sergio and Jose while eating an ice cream cone.	-Nicole says, "Hi boys> Mr. Page is giving out free ice cream. You should go get some."
-Nicole walks away.	

Credit

I would like to thank my daughter Zea for taking the time out of her busy schedule to help me edit and revise my work.

Notes:

www.ingramcontent.com/pod-product-compliance
Lightning Source LLC
LaVergne TN
LVHW061340060426
835511LV00014B/2041